Physic Mysteries
The Structure of Reality

A Journey into the Heart of Existence: The
Fabric of Reality: Unraveling the Secrets
of Physics

Kitchen Mage

The selling point

Take a deep dive into the universe with our ground-breaking book, "Unlock the Universe." Through the skillful integration of interactive multimedia elements with engrossing storytelling, this book provides an educational experience that goes beyond conventional textbooks. With the help of entertaining films, interactive simulations, interactive forums, and thought-provoking conversations, readers may explore the mysteries of physics like never before. Discovering the beauty, intricacy, and interconnection of the universe in a way that piques interest, awakens creativity, and fires a lifetime enthusiasm for scientific study is what "Unlock the Universe" offers everyone, whether they are lifelong learners, educators, or students.

The intended viewers

1. Pupils and instructors looking for a thorough grasp of physics principles.
2. The latest findings and theories in science pique the interest of readers.
3. philosophers who want to investigate the relationship between metaphysics and physics.
4. People find the implications of quantum physics on our understanding of reality to be fascinating.
5. Authors and producers of science fiction who draw inspiration from scientific ideas.
6. amateur cosmologists and astronomers who want to learn more about the universe.
7. Anyone curious about the nature of reality and the basic components of existence.
8. professionals who wish to improve their understanding of fundamental physics and work in disciplines like engineering, technology, and medicine.

9. Spiritual seekers are interested in making connections between existential and philosophical questions and scientific principles.
10. Book groups discussed challenging non-fiction literature.
11. Enthusiasts of science communication strive to communicate difficult scientific ideas in understandable ways.
12. anyone thinking about pursuing a profession or academic path in physics or similar fields.
13. Inquisitive minds of all ages want to deepen their knowledge of the natural world, from teenagers to pensioners.
14. Those who enjoy popular science books and videos and are curious to learn more about the wonders of the cosmos
15. Scientists and researchers looking for new ideas or inspiration for their physics research.
16. readers with an interest in the development of scientific ideas and the history of physics.

17. Discussion groups and philosophy clubs investigate how scientific discoveries affect our perception of reality.
18. Libraries and bookshops trying to provide their customers with a wide range of science fiction titles.
19. Online forums devoted to physics, cosmology, and associated subjects are searching for interesting articles to post and debate.

I. Overview of the Structure of Reality

- synopsis of the basic ideas in physics.
- the investigation into what makes reality what it is.
- A brief account of the major physics findings.

II. Investigating the Quantum Universe

- An overview of quantum mechanics.
- Particles have two natures: waves and particles.
- The ramifications of quantum entanglement.
- Different interpretations of quantum physics, including the Copenhagen interpretation.

III. Revealing Relativity's Mysteries

- Overview of general and special relativity.
- The curvature and nature of spacetime.
- contraction of length and dilation of time.
- Spacetime's fabric and black holes.

IV. Nature's Unified Forces

- the pursuit of a cohesive physics theory.
- Overview of particle physics' standard model.
- Strong nuclear force, weak nuclear force, electromagnetism, and gravity are the four fundamental forces.
- Other attempts at unification include supersymmetry and string theory.

V. Extending the Standard Model

- The mysteries of the universe include dark matter and dark energy.
- investigating the limits of particle physics.
- looking for physics outside of the conventional model.
- The properties of the multiverse and quantum gravity.

VI. Physics as Philosophy

- The consequences of modern physics for philosophy.
- The connection between metaphysics and physics.
- Quantum mechanics and consciousness: a relationship.
- discussions on the nature of reality, free will, and determinism.

VII. Utilizations and Consequences

- Useful applications of physics in engineering, technology, and medicine.
- Physics's effects on culture and society.
- ethical issues in technology creation and science study.
- Theories concerning the direction of physics and how it will influence human destiny.

VIII. Concluding Remarks: Accepting the Marvel of the Universe

- Thoughts about the universe's complexity and beauty.
- the continuous investigation into reality's structure.

- a challenge to the readers to keep delving into the secrets of existence and the miracles of physics.

IX. Uniting Spirituality and Science

- and investigating connections between ideas from science and spirituality.
- A scientific and spiritual perspective on the interconnectivity of all things.
- striking a balance between one's personal belief system and scientific understanding.
- The place of wonder and amazement in both spiritual and scientific investigation.

X. Useful Hints and Activities

- Deepening understanding through practical trials and cognitive exercises.

- exercises to assist readers in integrating physics ideas into their daily lives.
- Ideas for meditation and introspection derived from the laws of physics.
- Resources and recommendations for more reading to carry on the exploration process.

XI. Visionaries and Pioneers: Profiles

- biographical sketches of prominent intellectuals and physicists who have influenced our conception of reality.
- insights about their life philosophies and contributions to science.
- Narratives of tenacity, revelation, and inventiveness motivate readers to continue on their exploratory travels.

XII. Physics Education's Future

- advances in physics outreach and education.
- Techniques for helping a large audience understand difficult physics subjects.
- The significance of cultivating inquisitiveness and analytical reasoning in upcoming cohorts.
- Possibilities for international cooperation in physics education advancement involving scientists, educators, and science communicators.

XIII. Interactive Guide to Multimedia

- A digital platform or companion website that offers films, interactive simulations, and additional resources.

- forums where readers can collaborate and have discussions.
- extra materials, like study guides and worksheets that can be downloaded.
- updates and additions to keep readers interested and up to date on the most recent advancements in physics.

I. An Overview of Reality's Fabric

The most comprehensive endeavor made by humans to understand the complex web of reality is the study of physics, which is concerned with matter, energy, and the fundamental forces that control the universe. Fundamentally, physics explores the fundamental ideas that form the universe and our place in it to solve the mysteries of existence itself.

This introduction chapter provides readers with a basic overview of the fundamental concepts of physics and the pursuit of understanding the nature of reality, so serving as a gateway into the fascinating field of physics. With the help of physics, we can examine the underlying order and complexity of the universe, from the tiniest subatomic particles to the vastness of space. An overview of history serves as the starting point for this voyage into the fabric of reality, following in the footsteps of visionary thinkers whose ground-breaking discoveries have impacted our comprehension of the material universe. Every significant development in the history of physics, from the groundbreaking discoveries of Isaac Newton and his laws of motion to the twentieth-century quantum revolution, has pushed us one step closer to solving the universe's riddles.

We welcome readers to join us as we set out on this investigation to learn more about the fundamental ideas that underpin the cosmos and to consider the important issues that have captured the attention of people for ages. Let's explore the fabric of reality together and go on an exciting adventure of discovery that will sharpen our senses, pique our interests, and increase our appreciation for the wonders of the universe. We will come across ideas on our voyage that are beyond our comprehension, such as the strange phenomena of quantum mechanics and the profound ramifications of Einstein's theory of relativity. We will examine the sophisticated mathematical structures that support our comprehension of the cosmos as well as the scientific techniques employed to delve into the secrets of nature. However, the pursuit of understanding the fundamental nature of reality itself lies deeper than mathematics and experiments. What are time and space made of? What underlying factors govern how matter and energy behave? Lastly, and maybe most importantly, how do we fit into the overall scheme of things?

We will come across philosophical questions that have captivated intellectuals for millennia together with scientific concepts as we make our way through the complexities of physics. The study of physics invites us to consider the fundamental structure of reality and our role within it, from arguments concerning the nature of causality and determinism to meditations on the meaning of consciousness and free will. So let us set out on this journey of the mind in unbridled curiosity and with open minds. In the search for knowledge about the structure of reality, we set out on a journey that cuts over space and time, providing us with ever-deeper insights into the unfathomable mysteries of life.

II. Investigating the Quantum World

Few domains in the wide field of physics are as fascinating and mysterious as the quantum universe. This chapter provides an introduction to the field of quantum mechanics, exposing readers to the paradoxical occurrences that define this area of physics and casting doubt on our conventional view of the cosmos.

The idea of wave-particle duality, which shows that particles like electrons and photons can behave like both waves and particles, is at the core of quantum mechanics. This duality is the cornerstone of quantum theory and the basis for many of its most profound findings. It was initially postulated by trailblazing scientists like Louis de Broglie and experimentally verified by seminal experiments like the double-slit experiment.

The phenomenon of quantum entanglement, in which particles become correlated to the extent that their states are instantly influenced by one another, regardless of their distance from one another, is one of the most amazing applications of quantum mechanics. This phenomenon, which Einstein dubbed "spooky action at a distance," defies traditional ideas of locality and causation and has significant ramifications for how we conceptualize reality.

One of the most widely recognized interpretations of quantum physics, the Copenhagen interpretation, is encountered as we continue our exploration of the quantum domain. In this interpretation, particles exist in a superposition of states until they are measured, and the act of measuring is fundamental to deciding the result of quantum experiments. The Many-Worlds interpretation and the Pilot-Wave theory, among others, provide other explanations for the strange events of the quantum universe. Nevertheless, the Copenhagen interpretation is not without its detractors.

We welcome readers to journey with us into the quantum realm in this chapter, where the rules of classical physics no longer apply and reality is woven from the exotic and enigmatic threads of quantum mechanics. As we work to uncover the mysteries of the quantum cosmos, let's investigate the profound implications and mind-bending ideas of this intriguing field of physics together.

III. Revealing Relativity's Mysteries

Albert Einstein's revolutionary theories of relativity at the beginning of the 20th century fundamentally changed our understanding of the universe. This chapter introduces readers to special relativity and general relativity, the two foundations of contemporary physics that together offer a fundamental understanding of the nature of gravity, space, and time. Einstein's introduction of special relativity in 1905 dramatically changed our understanding of space and time by proving that they are not absolute but rather related components of a single phenomenon called spacetime. By special relativity, measurements of time and space can change based on the relative velocity of observers, even though the laws of physics apply to all observers moving in uniform motion. In addition, this theory explains phenomena like length contraction—where objects appear shorter when traveling at relativistic speeds—and time dilation—where time appears to pass more slowly for objects in motion relative to an observer.

Einstein developed general relativity in 1915, expanding on the ideas of special relativity to take gravity into account. According to this theory, mass and energy cause spacetime to curve, not huge things exerting force. This is how gravity is explained. General relativity states that spacetime is warped by huge objects like planets and stars, which causes other objects to travel through space on curved pathways. The gravitational time dilation felt close to black holes and the bending of light around big objects are phenomena caused by this curvature of spacetime.

Speaking of black holes, these mysterious cosmic objects are among general relativity's most intriguing predictions. Nothing can escape the gravitational pull of a black hole—a region of spacetime where gravity is so powerful that not even light can. Spacetime is so severely warped around a black hole's event horizon that traditional concepts of space and time completely collapse, resulting in strange events like singularities forming and time itself being distorted.

We urge readers to explore the world of relativity in this chapter, where mass, energy, and curvature interact to weave the fabric of spacetime. Let's investigate the tremendous effects of Einstein's theories on our comprehension of the universe and consider the enigmas surrounding black holes, gravitational waves, and spacetime itself.

We come across events that defy our most ingrained beliefs about the nature of reality and push the boundaries of our imagination as we continue to explore the secrets of relativity. The fundamental ideas of relativity provide a deep window into the fundamental makeup of the cosmos, explaining phenomena like the lengthening of time in the presence of gravity and the warping of spacetime around big objects. The existence of gravitational waves—ripples in spacetime that travel through the universe at the speed of light—is one of general relativity's most stunning consequences. Cataclysmic events like black hole collisions and neutron star mergers produce these waves, which carry information about the nature of the objects that produced them. Recent gravitational wave discoveries made by LIGO and Virgo have given us a fresh perspective on the cosmos and enabled us to investigate phenomena that were previously unobservable.

The most recent advancements in relativity will be discussed in this chapter, ranging from the discovery of gravitational waves to the continuous hunt for concepts like dark matter and dark energy. We will also look at how relativity has shaped our conception of the world at its grandest scales, including its composition, evolution, and ultimate destiny. Come along with us as we explore the depths of spacetime, where the complex interactions of matter, energy, and gravity weave the fabric of reality. Together, let's explore the complexities of relativity and take in the breathtaking beauty of the universe as shown by Einstein's ground-breaking theories.

IV. Nature's Unified Forces

Physicists have long sought a grand unified theory—a single framework that unifies all the fundamental forces of nature under a similar set of principles—in the quest to comprehend the fundamental rules that govern the universe. The search for such a theory is examined in this chapter, which also gives readers an overview of particle physics' standard model, which offers a thorough explanation of the fundamental particles and forces that comprise the universe. Particles are divided into two primary types according to the standard model of particle physics, which was created in the second half of the 20th century: bosons and fermions. Bosons, like photons and gluons, are the carriers of the fundamental forces, while fermions, which include quarks and leptons, are the building elements of matter. Numerous studies carried out in particle accelerators worldwide have validated this model, which elegantly describes a wide range of phenomena observed in particle physics experiments. But even with its achievements, there are still certain drawbacks to the conventional paradigm. Its failure to account for gravity, a force that general relativity describes but which is elusive at the quantum level, is one of its most obvious flaws. The quest for a single, cohesive theory that unifies the ideas of general relativity and quantum mechanics has been spurred by this disparity. In the pursuit of unification, supersymmetry and string theory are two potential directions. The proposal of supersymmetry suggests a symmetry between bosons and fermions, which may be able to address some of the unresolved problems in particle physics, including the existence of dark matter and the hierarchy problem. Contrarily, string theory postulates that the fundamental particles of the universe are microscopic, vibrating strings, whose various vibrational modes give rise to the various particles and forces seen in nature, rather than point-like particles.

We will look at these and other attempts at unification in this chapter, as well as the theoretical foundations that support our knowledge of the fundamental particles and forces that control the universe. Come explore the secrets of particle physics and the cutting edge of theoretical physics, where scientists from all over the world are still motivated and faced with challenges in their pursuit of a single, cohesive theory of nature.

V. Extending the Standard Model

Our knowledge of the enormous mysteries that exist outside the parameters of the standard model of particle physics has grown along with our comprehension of the fundamental forces and particles that control the cosmos. This chapter explores the frontiers of contemporary physics, delving into the intriguing possibilities of a multiverse, dark matter, and dark energy as we search for new physics outside the standard model.

In contemporary cosmology, two of the most puzzling mysteries are dark matter and dark energy. Dark matter, which makes up most of the universe's mass-energy content, is invisible to telescopes and has not been directly detected in lab studies. Its gravitational pull on galaxies and galaxy clusters suggests its existence, but its exact nature is still a mystery. On the other hand, dark energy is believed to be the cause of the universe's apparent acceleration of expansion, although its source and characteristics are yet unknown. When combined, dark matter and dark energy put our knowledge of the universe to the test and entice us to push the boundaries of observational and theoretical physics.

In particle physics, the hunt for new events outside of the standard model is at the forefront. To uncover new particles or interactions that defy existing theories, researchers perform experiments using particle accelerators like the Large Hadron Collider (LHC) to investigate the fundamental basis of matter and energy at ever-higher energies. Scientists are pushing the limits of our knowledge to comprehend the world on a deeper level. This includes the quest for supersymmetric particles as well as the investigation of extra dimensions and novel gauge symmetries. The search to reconcile the laws of general relativity with quantum mechanics is one of the most significant issues facing theoretical physicists. Even the most bright minds in the field have not been able to accomplish this task thus far. One of the great mysteries of contemporary physics is quantum gravity, which aims to combine the gravitational description of spacetime with the quantum description of particles. It holds the potential to reveal the basic mysteries of the cosmos.

The idea that there are multiple parallel worlds within a large cosmic landscape has become intriguing in recent years, raising the possibility that our universe is merely one of them. Though purely theoretical, the concept of a multiverse questions our standard understanding of reality and prompts us to consider the possibility that there are other worlds besides our own. We will venture into uncharted territory in this chapter as we investigate the enigmas of dark energy, dark matter, quantum gravity, and the alluring possibility of a multiverse. Come along with us as we explore the boundaries of contemporary physics, where the search for new physics beyond the standard model promises to shed light on the universe's darkest corners and broaden our understanding of it in almost unfathomable ways.

VI. Physics as Philosophy

The deeper our grasp of the universe has become through the application of contemporary physics, the more profound the philosophical implications of our scientific findings have become. This chapter explores the rich tapestry of philosophical inquiry that accompanies our investigation of the underlying laws of the universe, ranging from the complex interactions between metaphysics and physics to the deepest inquiries into the nature of reality itself, consciousness, determinism, and free will.

The topic of how our notion of reality is shaped by our scientific understanding of the cosmos is at the center of the philosophical debate surrounding modern physics. Every discovery, from the ground-breaking revelations of quantum physics to the mind-bending consequences of general relativity, calls into question our preconceived conceptions of matter, energy, space, and time and forces us to reevaluate how we see the essence of existence. Philosophers and physicists have long disagreed on the connection between metaphysics, a field of philosophy that addresses the essential nature of reality, and physics. While metaphysics explores deeper issues like the nature of consciousness, the ultimate essence of reality, and the nature of being itself, physics aims to identify the fundamental rules and principles that control the behavior of the physical universe. When combined, these two fields of study provide contrasting viewpoints on the essence of life and encourage us to investigate the limits of the metaphysical and the physical.

The investigation of how consciousness shapes our perception of the cosmos is among the most fascinating facets of contemporary physics. The behavior of particles in the world of quantum mechanics is largely determined by observation, a characteristic that has sparked conjecture over the place of consciousness in the quantum world. A more fundamental role for consciousness in the structure of reality itself is suggested by some physicists, while others maintain that consciousness is just an emergent feature of complicated physical systems. Some of the most lasting questions in philosophy and physics are those related to determinism, free choice, and the nature of reality. Philosophers and physicists debate the nature of causation, the presence of free will, and the predictability of the universe, ranging from the deterministic worldview of classical mechanics to the indeterminacy of quantum mechanics. These discussions call into question our conceptions of agency, accountability, and the essential character of reality. They also encourage us to reflect on the philosophical and scientific aspects of the riddles surrounding existence. We will examine these and other philosophical ramifications of contemporary physics in this chapter, exploring the complex relationship between philosophy and science as we seek to understand the fundamentals of reality. Join us as we set out on a voyage of intellectual discovery where the lines separating philosophy and physics are blurred and our pursuit of understanding pushes us to the brink of human knowledge.

VII. Utilizations and Consequences

The basic science of physics, which forms the basis of our comprehension of the natural world, has significant ramifications for engineering, technology, health, and society at large. This chapter looks at the real-world uses of physics in a variety of fields as well as the larger effects of physics on culture, society, and ethical issues in science and technology advancement. We will also make some conjectures about the direction that physics will take in the future and how it might affect the fate of humanity.

Uses of Physics in Real-World Applications
The modern world is shaped by innovation and is driven by the laws of physics, which have numerous applications in engineering, technology, and medicine. Physics has revolutionized the realm of electronics, opening the way for the information era and powering the gadgets that have become vital in our everyday

lives, from the invention of the transistor to the development of lasers and semiconductors. Physics is essential to medical imaging methods like MRIs, PET scans, and X-rays because they allow doctors to identify and treat a wide spectrum of illnesses with previously unheard-of precision. Modern cancer treatment methods like radiation therapy and proton therapy, which give patients more effective and minimally intrusive treatment alternatives, are also based on physics. Physics is used in engineering to design anything from cars and airplanes to skyscrapers and bridges, making sure that the constructions are strong, safe, and able to withstand environmental conditions. Superconductors for energy transmission and lightweight composites for aircraft are just two examples of sophisticated materials whose features have been developed thanks to the discipline of materials science, which mostly relies on physics.

Physics's effects on culture and society

Physics has a tremendous influence on society and culture that goes well beyond technology and medicine. Physics has revolutionized our view of the cosmos and our role within it, generating astonishment and igniting the human desire for knowledge and understanding from the scientific revolution to the modern period. From the portrayal of scientific subjects in visual art and music to the investigation of the cosmos in literature and film, the study of physics has also helped shape cultural narratives and artistic expressions. Scientists who have contributed to our understanding of the universe and who arouse curiosity and imagination, like Stephen Hawking and Albert Einstein, have become cultural icons.

Scientific Research and Technological Development: Ethical Issues

Although there is much hope for the betterment of humankind through the use of physics, there are significant ethical issues that need to be carefully explored. Physicists and engineers deal with difficult moral conundrums that need careful thought and deliberation, ranging from the moral implications of nuclear weapons and energy to worries about privacy and monitoring in the era of digital technology.

We must keep a close eye on the possible effects on society and the environment as we continue to push the limits of scientific understanding and technological innovation. Our decisions and actions should be guided by ethical concepts like transparency, accountability, and equity. This will help to ensure that the benefits of scientific development are distributed fairly and that any hazards are minimized as much as feasible. Conjectures Regarding the Prospects of Physics In terms of expanding human knowledge and talents as well as deepening our grasp of the cosmos, the field of physics has a bright future ahead of it. In their quest to solve cosmic riddles and unearth novel theories that will influence the trajectory of human history, physicists are exploring the quantum realm and looking for new physics beyond the standard model. Technological advances in areas like space exploration, fusion energy, and quantum computing could transform our knowledge of the cosmos and our capacity to use its resources for human benefit in the decades to come. Scientists and philosophers alike will continue to be inspired and challenged by the pursuit of a unifying theory of physics and the investigation of strange phenomena like dark matter, dark energy, and black holes, which will expand human knowledge and alter our perception of reality.

We have looked at the many uses and ramifications of physics in technology, health, society, and culture in this chapter. We have also discussed the moral issues that arise from scientific advancements and technological progress. We can only hypothesize as to how physics may influence human destiny in the future, but one thing is for sure: science will continue to advance and reveal the mysteries of the cosmos to future generations as long as people are driven by a desire for knowledge and exploration.

VIII. Concluding Remarks: Accepting the Marvel of the Universe

We are in awe of the beauty and intricacy of the universe around us as our trip through the domains of physics comes to an end. The wonders of the world, from the subatomic particles that dance in the quantum realm to the magnificent galaxies that adorn the cosmos, inspire us to reflect on our role in the grand scheme of things.

We have come upon the complex interactions between matter, energy, space, and time—the fundamental components of reality itself—during our exploration. We have marveled at spacetime's curvature, wrestled with the puzzles of quantum mechanics, and wondered about the multiverse and black holes. Nevertheless, despite everything we have learned, the cosmos still begs us to carry on with our investigation into its mysteries.

This is not where the voyage of discovery ends. We are urged to embrace the wonder of the cosmos and to continue our quest for understanding with open minds and curious hearts as readers, scientists, philosophers, and explorers. The pursuit of comprehending the essence of reality is a continuous expedition that surpasses temporal and spatial limitations and challenges us to delve into the most profound enigmas of life.

Thus, let us take with us the spirit of inquiry and sense of wonder that has led us on this voyage as we bid these pages goodnight. Knowing that every discovery brings us one step closer to discovering the mysteries of the cosmos, let us continue to investigate the miracles of physics and the riddles of existence.

The famous physicist Richard Feynman once said, "If you look at it right, the world is a dynamic mess of jiggling things." And when you enlarge it, everything is moving so much that you can scarcely see anything anymore. However, it turns out that you can see this amazing, enigmatic phenomenon occurring more clearly the more you magnify the globe." May we never stop appreciating how amazing and mysterious the cosmos is, and how we may magnify the globe.

IX. Uniting Spirituality and Science

Science and spirituality are two pillars of knowing in the enormous field of human research; each has a distinct viewpoint on the secrets of existence. This chapter examines the fertile ground where these two domains meet, encouraging readers to consider the interconnection of all things, look for harmony between scientific understanding and personal belief systems, and investigate the similarities between scientific notions and spiritual views. Discovering the mysteries of the cosmos and our role in it has always been a goal shared by scientists and spiritual searchers alike. Finding the fundamental truths underlying the fabric of reality is the same objective of both professions, despite the differences in their methodologies and approaches.

Finding similarities between scientific ideas and spiritual beliefs is one field of investigation. Several places of confluence challenge our conceptions of what it is to be alive, from the idea of interconnection in physics to the idea of global awareness in spirituality. Both spirituality and science provide insights into the essence of reality and our role in it, whether they are pursuing the quantum world or examining the depths of consciousness.

The realization that everything is interconnected—a notion that is profoundly meaningful from both a scientific and spiritual standpoint—lays the foundation of this investigation. Everything in the cosmos is interrelated, ranging from the tiniest subatomic particles to the enormous expanses of space, creating a web of interdependence. This understanding challenges us to accept the intrinsic oneness of all things and go beyond the bounds of personal identity.

A really rewarding and enlightening activity may be achieved when scientific understanding and personal belief systems are brought into harmony. We may develop a more comprehensive knowledge of the cosmos and our role in it by realizing the complementarity of scientific and spiritual discoveries. We can accept science and spirituality as complementary modes of knowing that provide distinct viewpoints on the same underlying reality, as opposed to seeing them as antagonistic or contradictory.

Both scientific research and spiritual investigation are guided by awe and amazement, which encourage us to reflect on life's mysteries and be in awe of the universe's beauty and complexity. The feeling of amazement and wonder that accompanies our investigation, whether we are looking into the depths of space or the depths of consciousness, urges us to go beyond the boundaries of our unique viewpoints and connect with something bigger than ourselves.

We encourage readers to join us in this chapter as we take them on an illuminating voyage of discovery where science and spirituality meet to reveal the universe's most profound truths. We can develop a deeper appreciation for the wonders of existence and our place within the cosmic tapestry of life by examining the similarities between scientific ideas and spiritual beliefs, thinking about the interconnectedness of all things, and finding harmony between scientific understanding and personal belief systems.

X. Useful Hints and Activities

This chapter contains useful tips and activities that will help readers comprehend physics ideas more fully, apply them to real-world situations, and develop personally via introspection and investigation. These activities are meant to captivate readers in a comprehensive investigation of the universe's marvels and stimulate lifelong learning and curiosity. They range from practical experiments to mental exercises and meditation suggestions.

Practical Tests and Mental Activities

Readers may experience ideas like momentum, energy, and wave behavior directly by participating in practical activities that make physics come to life. Building basic devices, doing physics experiments at the kitchen table, or investigating the characteristics of light and sound are just a few examples of the practical ways that these pursuits may foster a greater awareness and respect for nature.

Readers are encouraged to investigate their intuitions and views about the nature of reality, as well as the deeper implications of physics principles, through thought activities. These activities inspire readers to think critically and imaginatively about the wonders of the universe, from pondering the complexities of quantum physics to considering the interconnectivity of all things.

Activities for Daily Life

Discover the usefulness of scientific ideas in our day-to-day lives by applying physics concepts to commonplace scenarios and problems. These tasks aid readers in gaining a greater understanding of how physics shapes the world around them, from figuring out the forces at work in a game of pool to maximizing energy efficiency in home appliances.

Prompts for Meditation and Reflection

Use ideas like symmetry, harmony, and connection as guiding themes for meditation and introspection. Use physics principles as inspiration to develop mindfulness and reflection. These questions encourage readers to engage with the intricacy and beauty of the cosmos on a deeper level, whether they are thinking about the cycles of the natural world or the patterns of the stars.

Resources and Suggestions for More Reading Take a look at a carefully chosen list of books, articles, movies, and other resources to carry on your investigation after you finish reading this book. These suggestions provide a multitude of avenues for more education and exploration, ranging from cutting-edge research and popular scientific publications to historic classics written by trailblazing physicists.

In this chapter, we encourage readers to interact with physics as a means of inspiration, wonder, and self-improvement, in addition to as a means of intellectual inquiry. Using practical applications, intellectual challenges, experiential experiments, and contemplations on the intricacy and beauty of the cosmos, we aim to enable readers to carry on their quest for knowledge and understanding long after they have concluded perusing these pages.

XI. Visionaries and Pioneers: Profiles

This chapter explores the lives and achievements of prominent philosophers and scientists whose ground-breaking work has altered the trajectory of human history and impacted our knowledge of reality. We hope to encourage readers on their paths of inquiry and discovery via biographical profiles, an examination of their life philosophies, and insights into their contributions to physics.

1. Einstein, Albert (1879–1955)

Albert Einstein, one of the most famous scientists in history, transformed our knowledge of the universe via his theories of relativity and his deep grasp of the nature of gravity, space, and time. Through his unwavering search for knowledge and his dedication to challenging accepted beliefs, Einstein upended the fundamental tenets of classical physics and established the framework for contemporary theoretical physics.

2. The late Marie Curie (1867–1934)

Marie Curie was the first woman to receive the Nobel Prize and the first person to win the prize in two separate scientific domains. She was a pioneer in the realm of radioactivity and her revolutionary study opened the way for other discoveries in physics and medicine. Curie encouraged generations of scientists to follow their passions and push the envelope of what is possible via her unrelenting dedication to scientific inquiry and her unyielding commitment to furthering human knowledge.

3. Feynman, Richard (1918– 1988)

Richard Feynman, who won a Nobel Prize for his work on quantum electrodynamics, was a well-regarded scientist, humorist, and insatiable curiosity. He made significant advances to our knowledge of quantum mechanics and particle physics. Feynman encouraged numerous students and readers to embrace the wonder of the cosmos and approach scientific research with an open mind and a fun spirit through his captivating lectures, colorful tales, and contagious passion for science.

4. 1942-born Stephen Hawking

Stephen Hawking was a visionary theoretical physicist and cosmologist who produced important findings concerning the nature of black holes, the universe's beginnings, and the basic laws governing it. Even though he had severe physical difficulties as a result of his ALS, Hawking's unwavering spirit, keen mind, and unquenchable curiosity helped him rise to prominence as one of the most important scientists of the modern era and encouraged millions of people to ponder life's greatest mysteries.

We learn about the tenacity, creativity, and discovery that have reshaped human history and molded our perception of reality via the accounts of these and other visionaries and pioneers. Their lives are examples of the strength of human ingenuity, curiosity, and perseverance; they encourage us to rise to the difficulties of exploration and pursue a better knowledge of the universe and our role in it.

XII. Physics Education's Future

This chapter examines how physics education is changing and considers innovative and cooperative teaching, outreach, and communication strategies. We examine how educators, scientists, and science communicators can collaborate to make complex physics concepts accessible to a wide audience and inspire the next generation of scientists and thinkers. From creative teaching strategies to the significance of fostering curiosity and critical thinking.

1. New Approaches to Physics Outreach and Education

Technological and pedagogical developments present stimulating prospects for physics education innovation. Teachers have access to a multitude of tools and resources that can improve learning and engagement for students of all ages and backgrounds, from interactive simulations and virtual laboratories to online courses and educational applications.

Furthermore, outreach programs like scientific fairs, open talks, and interactive exhibits give scientists and educators a chance to engage with the general public and pique their curiosity about physics.

2. Creating Accessible Concepts in Complex Physics

Complex physics ideas can frequently appear intimidating or unapproachable to pupils, especially those who may not have a solid foundation in science or arithmetic. Nonetheless, educators may make even the most complex concepts interesting and clear with the correct methodology and materials. Through the use of analogies, practical examples, and visualizations, instructors may assist students in understanding difficult ideas and cultivate a greater appreciation for the elegance and beauty of physics.

3. Developing Inquisitiveness and Analytical Skills

The development of students' curiosity and critical thinking abilities is the main objective of physics education. Teachers can enable students to challenge preconceptions, pose questions, and engage in creative thinking about their surroundings by promoting inquiry, experimentation, and discovery. Using experiential learning, group assignments, and flexible research, pupils may cultivate the abilities and perspective required for triumph in physics and other subjects.

4. Cooperation and Joint Venture

Collaboration and partnership between educators, scientists, and scientific communicators are necessary to advance physics education. A more inclusive and accessible learning environment for kids from all backgrounds may be created by stakeholders cooperating to establish creative outreach initiatives, produce cutting-edge curricula, and exchange best practices. Furthermore, collaborations between government, business, and academia can aid in closing the knowledge gap between theory and practice by giving students access to real-world experiences and chances for career exploration and growth.

In summary, there is a bright and promising future for physics education. To excite the next generation of physicists and thinkers and guarantee that the wonders of the cosmos remain accessible to all, educators, scientists, and scientific communicators must embrace innovation, nurture curiosity, and critical thinking, and collaborate and cooperate. We can influence the direction of physics education and provide students with the tools they need to pursue their curiosity and desire to learn about the universe with passion and inventiveness if we work together and are committed to quality.

XIII. Interactive Guide to Multimedia

We are happy to provide an interactive multimedia companion to go along with this book. This companion website or digital platform is intended to improve your learning process and provide you with more tools to help you grasp physics principles on a deeper level. supplementary website or online resource Examine interactive movies, simulations, and other resources that make physics ideas more relatable and offer practical learning opportunities. These interactive tools, which range from virtual labs to dynamic visualizations, provide an entertaining method to study and experiment with important physics concepts.

Discussion and Collaboration Forums
Use discussion and cooperation areas to establish connections with instructors and other readers. Engage in dialogue, ask questions, and share your thoughts with a group of students

who are enthusiastic about scientific education and physics. Take part in thought-provoking discussions, ask for guidance, and work together on projects and activities to increase your comprehension and appreciation of physics.

Extra Materials

Get printable study guides, worksheets, and other materials to enhance your education and help you remember important ideas. These tools are a great way to help you understand the content and meet your learning objectives, whether you're seeking more practice problems, review materials, or study aids.

Revisions and Additions

Regular updates and additions to the interactive multimedia companion will keep you up to speed on the most recent advancements in physics. Explore fresh movies, simulations, and other materials created to keep you interested and up to date on the most recent advancements and discoveries in the subject. You can discover new and fascinating stuff to study and appreciate, whether it's updates on cutting-edge research or additions to previously published material.

In conclusion, this book's content is enhanced and complemented by the dynamic and immersive learning experience provided by the interactive multimedia companion. All on one handy platform: explore interactive simulations, participate in reader conversations, access extra resources, and keep up to speed on the most recent advancements in physics. We cordially welcome you to embark on this interactive voyage of inquiry and discovery with us as we study, learn about, and uncover the mysteries of physics.